Tour of
Insanity

Fantastic Things to do with a Dead
Body: Planning Your Life After Death

KELLY MITCHELL

authorHOUSE

AuthorHouse™
1663 Liberty Drive
Bloomington, IN 47403
www.authorhouse.com
Phone: 833-262-8899

Published by AuthorHouse 10/07/2021

ISBN: 978-1-6655-4072-8 (sc)
ISBN: 978-1-6655-4056-8 (hc)
ISBN: 978-1-6655-4088-9 (e)

Library of Congress Control Number: 2021920805

CONTENTS

PREAMBLE

When my son was 9-years-old, I found him standing in front of our fireplace staring up at our beloved dog's urn and pondering over the profound questions of life and death, or so I thought. As my middle child, he is analytical, advanced in his mental age and has a dark, sharp blade of a mind.

He looked at me for a moment, his eyes full of questions. The innocent, curious look on his face was always disturbing. I imagine the accused at the Salem witch trials felt less threatened than I did when this look was on his face.

My boy has a long, tumultuous history of making my single friends feel uncomfortable by telling them all about why they are single. He is the type of kid that when you engage in what seems like simple conversation, unexpectedly nose dives and takes a drastic complicated turn - spiraling off the plot.

These are the types of conversations in which you walk away

questioning your once unshakeable convictions and wonder how the hell that happened. The boy then shows his frustration at your weakness to adequately explain things to his satisfaction - it's humbling.

I braced myself for the impact of the uncomfortable conversation to come, and he didn't disappoint. I looked up at the urn and looked at the boy, raising my eyebrows in an invitation to fire away - and we were off.

Boy: "That's Toby, isn't it."

Me: "Yes." (*That's right momma, short and simple. So far, so good*).

Boy: "Why did you put her in a vase?"

Me: (*OK. There's the fishing question of entrapment. Think carefully, woman*). "Because we love her and want to keep her with us always. This way, we can admire the vase and reminisce about the fun times that make us smile and warm our hearts."

He gave pause as he processed what I just said. I saw the little wheels turning, and I thought to myself: *You did good, mom. You delivered a heart-felt, thoughtful answer with an endearing sentiment-mastery level. No way he can turn this around and make me the asshole.* I was wrong.

My inner thoughts lied to me. On his face, I saw the wind-up for the pitch. His lips pursed and his keen eyes settled in a raptor-gaze on mine-I could tell he was onto something. I slapped my inner monologue for acting like a noob. The boy shook his head in disapproval at my inadequate answer, pointed to the urn containing our dead dog, inhaled and blasted me with:

"So, that's it then? We live, work, and play in our lives, and for what? We end up in a box on someone's fireplace? Our life story in a vase ending up at a garage sale for 25 cents? Or, accidentally broken when the rest of our family dies, and no one is left to care for us? I know it's just our bones and not our soul, but that's messed up and seems selfish. I expected more."

And there it was—the curveball. STRIKE! *Expected more?!?*

He waited for my response and huffed at me when my shoulders slumped in defeat and I didn't give him one. This conversation was above my paygrade at the moment, and I thought it best to wait till I had an informed answer for him. This topic was going to take some research on my part to catch up. He spun on his heel and walked away, throwing his tiny hands in the air from frustration and disappointment, an action that deflated my '*bravo mom*' balloon (again).

I realized the boy wasn't wrong. He was alarmingly correct. Here I was, rethinking my philosophies and questioning my beliefs - damn this kid. Ending up in a vase on a fireplace wasn't appealing to me either. On the other hand, the thought of being underground in a casket was equally chilling because I am claustrophobic.

The question became: *What is appealing and would make me smile thinking about it?* And so, my quest for an answer began. Because of this abrupt and startling conversation with my son, I came across several alternatives and found a solution that excites me much more than a vase on a fireplace. I also found several current practices that were much worse.

That boy of mine can be infuriating beyond words. Still, in his forced self-reflection of my life choices, I always find something to be improved upon. In this case, how to be immortal in style, commemorate loved ones, preserve our history, and help save our planet.

This book is dedicated to my handsome, bold, evil-genius middle child as a reminder to stay curious and push people out of their comfort zones into something better.

CHAPTER 1

Ashes to Ashes

The flipside about being alive is that someday you will be dead. As we get older, our death preoccupies a substantially more significant portion of our thoughts - or when some type of apocalyptic event rocks our world (talking to you, COVID). Death has been an intimate part of life for hundreds of thousands of years. About 130,000 years, give or take a decade.

The belief of life after death has coupled the life/death coin for just as long. Ancient Neanderthals buried their dead with tools and belongings they may need in the afterlife. Neanderthals are considered the first human species to begin burying the dead. The <u>reasons</u> for ancient civilizations burying dead vary, but most circle the wagons around:

- Prevent the odor of decay
- Respect for the dead (scavengers could destroy a corpse if not protected)
- Give family closure
- Prevent the psychological trauma of witnessing decomposition
- Ritualistic (burial must happen to reach the afterlife or participate in the circle of life)
- Religious (many religions require specific customs to be laid to rest)

Not necessarily in that order. Burying isn't narrowed to only soil or consecrated ground venues. People were buried in mounds of earth, caves, and temples before the prominence of underground burials complete with a stone marker took precedence in most cultures. Cremation remains more prevalent in some places like India. Cremation is <u>mandatory</u> in Japan.

Religion and death go together like peas and carrots. Any belief you choose (even atheism) will influence how you live your life, how you perceive death, and how you want your belongings - including your remains - to be handled. To understand where we are going, it is essential to know where we have been. Since this is a Tour of Insanity Book and religious beliefs (or lack thereof) ties closely to how we live and die - a brief exploration of the history of disposing of bodies is required. Full disclosure, this is by no means a deep dive into the dirt.

Hindu - The Oldest Religion & Reincarnation

Before Hinduism took hold, Cannibalism was not out of the ordinary. Those who didn't feast on the dead buried them in mounds. Some Indian cultures also exposed <u>bodies</u> on the tops of buildings, called Towers of Silence, where vultures could eat their fill. Hinduism changed the burial landscape with its religion and landed about a billion followers.

The fundamental belief in Hinduism is Sanatana Dharma, better known as <u>Universal Law</u>. It sounds confining, but really, all it means is each person walks their path in a continuous cycle of death and rebirth (reincarnation)- trying to do better in the next life to reach Mukti (enlightenment). If you are devout enough to gain the insight and virtue of reaching Mukti in one of your life cycles, you are free from the reincarnation cycle.

Since you live and die hundreds to thousands of times, there is not much emphasis placed on the human body after death because the soul has departed. The body is viewed as a prison for your soul until you gain the wisdom to free yourself from the cycle of navigating the mortal world. Death is not permanent; the soul will occupy a new body.

In ancient culture, the body would be perfumed, adorned in flowers, and burned. The ashes would be thrown into the Ganges River (water is sacred). The burning releases any lingering ties so the soul is free to inhabit another body or move toward Mukti.

Before it was outlawed in 1829, Sutee (meaning 'faithful wife') was practiced. Sutee was when a widow would be burned on her husband's funeral pyre to join him in the next life, which helped her husband and herself toward Mukti. How viking, right? Babies, children, and saints were usually not put on pyres and were buried. It is believed those souls are pure and didn't fully attach to the body so burning wasn't necessary.

Very little has changed in the way of funeral rites for devout Hindus. In India, bodies are cremated along the Ganges River during a month-long set of traditions that prepare the soul to move on. Hindus located in America will either be cremated here or send their bodies to India. It is common for families to observe the cremation. Cremation may be the answer for Hinduism, and other religions, because the body is not important. But, what if you have been taught that it is?

Abrahamic or the '*No Burner*' Religions

Judaism, Christianity, and Islam started in the Middle East and, believe it or not, are <u>not</u> the first monotheist religions. When polytheism was all the rage, Egypt practiced Antenism for about 20 years. Antenism worshiped the sun, and it is believed in some circles that Judaism was born from this monotheistic religion - I wasn't there, so that last part is hearsay.

We know Christianity was born from the Jewish tradition, and Islam was formulated from a combination of Christianity and Judaism. The webs we weave, right? So, how did the religious DNA start splitting? The answer is Abraham--well him <u>and</u> the bloody Romans.

All three religions recognize Abraham as the first Jew that made a covenant with God. This makes him the first prophet in all three religions = Abrahamic. At the time, Romans had scattered Jewish society all over the map due to insurrections to Roman authority and occupation. When the State of Israel was founded in 1948, the scattered communities migrated to Israel. Unsurprisingly, they were not unified when reunited. Naturally, different geographic locales led to a variance in beliefs, languages, cultures, food, customs, and *religions*.

If you think about it, there are always different levels of devotion and belief within a religion. You have those who go to church, those who listen on apps, those who go on Christmas Eve or Easter, those who don't go (but make sure their kids do) -- there are many different relationships with one's chosen religion. <u>Global Connections</u> on PBS put it this way:

Orthodox Jews believe that Jewish law is unchanging and mandatory. Conservative Jews argue that God's laws change and evolve over time. Reform and Reconstructionist Jews believe that these laws are merely guidelines that individuals can choose to follow or not. In addition, there are many Jews in the United States who are secular or atheist. For them, their Judaism is a culture rather than a religion.

Putting it in the simplest of terms:

- Abraham founded Judaism
- Jesus founded Christianity
- Mohammad founded Islam

All three believe in the judgment, fire & brimstone, along with the gentle, forgiving, pacifist nature of the dual-sided divine they worship - which is the reason why all three frown upon cremation. If scripture is taken literally, there is a no-burn policy often accompanied by specific burial rites. The main reasons:

- Judaism = we are made in God's likeness
- Christianity = Jesus rose from the dead in his same body

- Islam = cremation does not fit with the respect and dignity of a dead body

Burials are the preferred means of laying a body to rest in these religions. To back up a bit to the patriarch of the jumping point for these religions - if you don't know the story of Abraham, we should skim that real quick, it's important.

In the book of Genesis, Abraham left Mesopotamia to become the father of a new nation per God's will. God promised him that his *'seed'* would inherit the land, so Abraham followed God's orders without question. One of God's requests was for Abraham to sacrifice his son, Isaac. Just as Abraham was to deliver on this request, God put the kibosh on it. God advised Abraham it was a test of faith, sparing Isaac.

Now, who the actual *'seed'* is, is seen differently in the three religions depending on which genealogy you put stock in. For Judaism, the belief is they are descended from Abraham's son Isaac. In Christianity, Jesus is traced to Isaac, and the near-sacrifice of Isaac is believed a foreshadowing of the crucifixion. In Islam, Abraham had a first-born son from another mother. His name was Ishmael, and Muhammad is *his* descendent.

Abraham is best known for the near-sacrifice of his son, nearly ending the descendent line. So, is the *'seed'* that of Isaac or that of Ishmael? I think the dispute could be ended if we just say it is the seed of Abraham - and put a pin in it. With all of this talk about sacrifice, we should probably look at some ways Aztecs disposed of a body.

Aztec Ritual Sacrifices

Evidence of sacrifices has been found in numerous cultures around the world. The depiction of rituals and grand events leading up to the ultimate sacrifice to appease an angry god - or calm a giant guerilla - has dazzled the Hollywood screen for decades.

Societies have frowned upon sacrifice for a while now since it fits within the definition of *'murder,'* so they are a rare sight to behold in real life. However, we will touch on a few civilizations where a sacrifice was a way of life.

Aztec life rotated around the capital of Tenochtitlan. Tenochtitlan was an epicenter of activity and socialization. Standing ominously in the capital was the Templo Mayor sitting snugly between two girthy towers constructed of human skulls. If that didn't send enough of a 'we will kill you' vibe, thousands of more skulls were impaled on racks as a screaming centerpiece. This layout was the first thing the Spanish conquistadors' saw of the Aztecs. Good first impression, right? One would think this would draw suspicion and make it impossible to relax much less eat a meal. To be blunt, you didn't know who you would be eating.

In 1521, conquistadors told tales of ceremonies they watched. Aztec priests with obsidian blades would plunge them deep into a victim's chest and hold up the still-beating heart to the gods. Then, after the heart stopped, the body would be tossed unceremoniously down the steps of the Templo Mayor, tumbling awkwardly down the decline. The people would watch and cheer. I would cheer too as long as it wasn't me.

Believing these tales to be rumors and false justification for the murder of Emporer Moctezuma, no one paid much attention other than the usual Ooos and Ahhhs that come with such stories. That is, until 2015, when archaeologists started working on an excavation site in Mexico City - where the Templo Mayor stood. Low and behold, they uncovered two skull towers, fully equipped with skull racks, and evidence of widespread human sacrifice. The count came to about 80,400.

Human sacrifice was a serious, intimately spiritual thing to the Aztecs and a matter of *survival*. Therefore, sacrifices were made throughout the year on specific calendar dates as preventative measures. Sacrifices were used to reverse adverse events (famine, drought), and dedicate temples, among other things. If sacrifices were not made, if the Aztecs did not feed the sun god human hearts and blood, the world would go into darkness and end.

A little side play to this belief was some serious Aztec intimidation flex to other cultures. This flex helped expand the Aztec empire. If you could sacrifice other people, not your own, that was a bonus. Captives and prisoners of war were the first to give their hearts to the sun. Thus, wars proved to be quite handy and a way to keep the sacrificial balance - I mean, sooner or later you would run out of Aztecs. Sacrifices of outsiders are supported through DNA tests conducted at Templo Mayor.

Human sacrifice is discouraged in today's complex, sophisticated societies. We are above that kind of brutal showcase. Are we, though? The hard truth is ritual human sacrifice *coincides* with the birth of a complex, sophisticated society. Human sacrifices are how social stratification is created and enforced. So, what is a more effective method than intimidating outsiders and your people? Two birds - one altar.

Human sacrifice is peppered in gladiator battles, mass burials of servants with Egyptian pharaohs and Chinese kings, burning of wives after the husband has passed, the bible, public executions - I mean, it's good to be the king, right? But, unfortunately, not all sacrifices are to a sun god. It's a sick game of thrones—one of which being the Aztec and Tlaxcalan *Flower Wars*.

Flower Wars were ceremonial battles, kind of like Survivor with a twist and entirely different prize. In these games, the two cultures would collect, not kill, as many enemies through combat as they could. The more people you catch, the more your social currency is elevated throughout your society. If the enemy captured you, you would have your heart cut out and participate in some ritual cannibalistic barbeque in your honor - while also being the main course.

In some cases, if you were a particularly sought-after delicacy, your head would be removed and what was left of your body given to high society members- in parts. Long live social gifting. Cannibalism is a hell of a way to dispose of a body. But, unfortunately, cannibalism is the gift that keeps on giving.

Cannibalism with the Fore

America's innocent *Leave It to Beaver* mindset was abruptly shattered in the 1950s when Australian gold prospectors had the happy accident of stumbling across an 11,000 member tribe called the Fore in Papua, New Guinea. This was not your usual tribe. At least 200 members a year were shivering, trembling, and subsequently dieing of an unknown cause. The shivering was the precursor to death. Once shaking started, death came swiftly. The Shivering---great title for a horror flick.

The infected <u>Fore</u> member would lose control of their limbs and would have trouble walking. They would also experience a loss of control over emotions. Think 'The Joker' from Batman-- the Fore even called it 'laughing death'. A year after losing motor control and emotions, the inflicted would become invalids who could not care for themselves.

These prospectors that found the Fore were concerned and reported this strange condition to scientists for fear of contagion. The Fore couldn't explain the cause either and considered the disease a part of life. What do you do when you don't understand something? You label it as *magic*, which is exactly what the Fore did. The condition mainly affected women and children. It is tough to procreate when you have no women, which led to the Fore being a dying breed. They were well aware of the threat of extinction looming and graciously invited researchers into their society to try and figure it out.

Researchers swooped in like knights in shining lab coats and spectacles, expediently checking off lists of demise starting with any kind of contaminant possible in the area and genetics. By this time, we had crossed into the 1960s, and a genetic specialist named Lindenbaum ruled out genetics. Instead, she followed the migration of the infliction and developed an absurd theory that ended up being spot on. The culprit that was sending the Fore into extinction was: Funeral rituals.

When people in the tribe died, the deceased would become the main course for consumption as a show of love and grief. The meal was prepared with love. Now, why this ritual didn't pique the interest of other researchers *before* Lindenbaum as a possible cause of the infliction is beyond me. It's not like Americans had seen anything like this before, so eating humans should have raised a few red flags and a bit of scrutiny. However, I imagine, if you pretend it doesn't exist--it doesn't.

The Fore practiced cannibalism because they didn't want the deceased family member eaten by maggots or insects and believed eating the recently deceased was more respectful. If you put it that way, I can kind of cosign the sentiment. The brain would be removed and mixed with ferns and fire-roasted with the body. Everything was eaten except the gallbladder. And everything was consumed by the _women_. This is why women were the majority inflicted with the disease.

If you are a mom, you know nothing you have is yours; it belongs to your child. In the spirit of mommyhood, and to quiet a child, mothers would pass a plate of fire-roasted relative to their greedy little hands. This is why children were also inflicted with the disease.

Why women? Women were believed to have the power of taming the evil spirits that come with a dead body. Women do have that whole nurturing thing that neutralizes and soothes a tantrum-throwing toddler. This skill apparently works on the dead too. At any rate, Lindenbaum's theory was so twisted it took a great deal of effort to get a biologist to set foot in New Guinea under those pretenses to prove the theory. When a biologist finally agreed, the US National Institute of Health injected infected Fore brains into chimpanzees and won the Nobel Prize for identifying the *Kuru* virus or *slow* virus.

The irony is, it's not really a virus. There is a specific agent in the human body that survives fire-roasting, boiling, and technically wasn't **alive**, to begin with. It is a *protein*. What's funny is I tell my kids, *If it has a mother, it is a protein*. I didn't know how right I was. This human protein which causes Kuro; we will call it Yoda - performed a Jedi mind-trick on other proteins in the human body. Other proteins started to mimic Yoda. Let's break that down for the non-Star Wars geeks by talking about zombies.

How do you get zombies? Take a zombie cell, bite a human cell and turn it into a zombie cell. The zombie cells bite more human cells and turn them into zombie cells. The zombie math kung-fu is *strong* with Kuru. Zombie infection is fast and furious. So too is Yoda - our Kuru virus protein. It works just like zombie cells. This makes a zombie apocalypse plausible, no? I'm enjoying my horror flicks so much more knowing this.

If you think that is scary, Kuru isn't the only thing you can get from eating dead bodies. Creutzfeldt-Jakob Disease is a degenerative neurological disorder like Kuru that can develop. Interestingly, per the Center for Disease Control & Prevention (CDC), one million people in the US have it. Read that again. How does that happen?

Once the Fore found out what was causing the disease, they stopped eating their loved ones around 1966. So it was a bit strange that cases still popped up; how could this be? It turns out the zombie cells, known as *prions,* take **decades** to show their effects. There are carriers. Per Michael Alpers at Curtin University in Australia, the last known case died in 2009. Now, let's blow your mind.

Remember mad cow disease? The CDC has repeatedly found people who have developed Creutzfeldt-Jakob Disease, the Kuru twin, in people that have eaten infected cattle. Scientists have not ruled out the possibility of becoming infected by ingesting other species. That's a species other than humans. Right now, the focus is on deer - for all you country boys out there.

Some deer and elk have developed Chronic Wasting Disease at an accelerated rate. Chronic Wasting Disease is another deal like Kuru. Chronic means this disease leads deer and elk to starve themselves to death. The exciting thing here is the zombie cells are prominent all over the deer and elk's body, not just the brain. The zombie cells show up in saliva, feces, and urine. That means zombie cells live in and *outside* the body. Remember, these zombie cells are not alive. That means a pile of poop is a ticking time bomb.

Hunters in Colorado and Wyoming probably don't know it, but the CDC and public health authorities monitor them watching for signs of zombie cells. Remember, they can take a **decade** to show up. We all have tiny zombie cells that ride alongside our normal cells. This is for real.

There is a Jekyll & Hyde in all of us. The balance of the brutality spectrum. Humans have always been walking contradictions: ying/yang, angel/devil, zombie cell/healthy cell. If we can't eat each other, there is a matter of storage after death to contend with. The question is, what are we going to be storing in your eternal resting place. When you are dealing with storage for an eternity, things can go wrong without a crystal ball telling the future.

CHAPTER 2

Eternity Wasn't As Long As I Thought

Planning for your eternal resting place may sound easy, but the reality is nothing lasts forever. There are nearly 8 <u>billion</u> people on this planet. Millions of people die every year (half a billion in the next decade), with the majority being cremated or buried. For every person breathing, there are **15** that are buried beneath the soil. The population above and below the earth takes resources and renders some resources useless. There is only so much room to cram people into the ground, and the business of death is expensive and taxing on the environment. We will get into the environmental toll a little later.

Cemeteries are often thought to be a forever kind of thing, the epitome of an infinity symbol - they are not. Cemeteries are businesses and can go under (pun intended), be relocated, go *missing,* or run out of space. A historic Brooklyn <u>cemetery</u>, a place of rest for many influential people, will run out of room in less than a decade. The notable Arlington National Cemetery will be full up mid-century. Other cemeteries are feeling the upheaval including Alaska and Florida. San Francisco stopped allowing new burials, and Seattle is close to following suit. This is just in America. America is a *young* country.

The island of England will have to find alternative means in the next 20 years. Beijing, China, has been full since 2016. In Africa, the fast-rising cities' infrastructure has no means to support burial spaces. Spain and Greece rent a temporary crypt for their loved one above ground. Usually, a few years' time is allowed for the body to decompose before the remains are moved to a communal burial ground, and the space is cleaned for a new rental.

Surprisingly, the 'moving after decomposing' or *reusing* of graves is a practice that America embraced. Americans had brought this practice with them from England. The common folk would be moved to common graves or storehouses after decomposition. Reusing graves is popular in Belgium, Portugal, Vancouver (BC), Germany, and Singapore. Reusing graves may be something that comes back in style in America. Why? In America, the urban cities boomed over where cemeteries used to be or were planned to be. Do you think your Homeowners Association (HOA) would want a graveyard next to the dog park? There is a stigma. Other countries are reaching a bit higher.

Asia and Latin America have put their stock in space-efficient vaults. The Memorial Necropole Ecumenica is 14 stories and holds tens of thousands of bodies in small vaults stacked on top of each other. Kind of like a giant safety deposit box or absurd game of Jenga. Taiwan has the True Dragon Tower, which is 20 stories and holds 400,000 people's ashes (#trending). Europe is taking notice of the efficient death cubicles with great interest. Paris went so far as to win the eVolo Skyscraper Competition with a towering vertical cemetery that included an impressive skylight, a circling ramp to visit loved ones, and astounding views.

Not wanting to be outdone, Norway started work on a honeycomb skyscraper cemetery poised to be the tallest building ever built. It would include a crane to lift coffins to dizzying heights. Hong Kong is not going up - but out.

Another option for the future: floating cemeteries. In Hong Kong, where thousands of residents already wait years for space in a public columbarium, the design firm Bread Studio developed a concept for a floating columbarium "island" called Floating Eternity. The island would dock to the mainland for ancestral worship holidays and then be sent back out to sea.

Does this mean that any of these things are built to last for an eternity and not be lost forever? No. The innovations are incredible, marvelous, unbelievable, national treasures, maybe even wonders of the world? But, so was the Taj Mahal.

Taj Mahal Travesty

The Taj Mahal was the beautiful, elaborate love letter from the Mogul emperor Shah Jahan to his favorite wife, Mumtaz Mahal, in 1632. She died during the birth of their 14th child. Now that's love - 14 ?! He built the Taj to last forever and stand the test of time to securely house his wife's remains. The Taj <u>mausoleum</u> is in India and is under both political and environmental threats. The destruction may have rendered the building too far gone to repair, and the consequence is a loss of an exquisite gift of architectural might to the world. Love may be eternal, but structures and political climates aren't.

The Taj Mahal is known for its pristine white marble, optical illusions, and supreme architecture, including a mosque, guest house, and formal gardens. Back in the day, the construction was around 32 million rupees - that's $200 million American, folks. Over 20,000 artists worked on the structure, and over 1,000 elephants transported the materials. This was the quintessential community project. Indeed, with such great care, time, and effort poured into the labor of love, someone would see to its preservation, right?

Wrong (image from newsbharati.com). The structural integrity of this wonder of the world is under scrutiny. What was once a white, shining beacon to the world is cracking and disintegrating at an alarming rate. The minarets are leaning precariously. The water level has declined in the Yamuna River basin causing the foundation to rot. It is incredible the mausoleum still stands as it was predicted to collapse in 2016.

Not only is the river declining at 5 ft per year, but the pollution going into the river has increased significantly due to the Mathura Oil Refinery, untreated sewage, and chemical run-off from Delhi. When you couple the concentrated water pollution with the air contamination, you get severe acid rain licking at the marble and turning it yellow. Emissions of Nitrogen Oxides are suffocating the Taj Mahal despite adopted emission standards.

The condition of the river attracts Goeldichrionomus insects and entices them to breed and infest the walls of the Taj Mahal. These bugs leave greenish-black patches over the once intricate floral designs. President Bill Clinton said it best:

Pollution has done what 350 years of wars, invasions, and natural disasters have failed to do and have begun to mar the magnificent walls of the Taj Mahal.

Pollution isn't the only thing threatening the Taj Mahal. The Taj Mahal is a Muslim-built structure sitting in a predominantly Hindu country. Some see the Taj Mahal as a blight on their land and not a representation of Indian culture, including many members of the government entities that would be responsible for the Taj's care. Terrorist threats to destroy the monument have occurred.

The government will be upping its security levels in 2022 at the Taj Mahal. The increase in security is primarily a credit to a petition in <u>2018</u> from the people and India's Supreme Court ruling in 2021 to restore the Taj or demolish it. As a result, the admission ticket went up 400%, and efforts are underway to save it; it remains to be seen whether they are too late. COVID tends to cut down tourism, but the crime and pollution around the Taj had tourist numbers plummeting long before COVID. If a majestic wonder like this can be threatened, what about our personal final wishes being carried out? Surely there are too many working parts in the Taj - a small personal wish is different and easily managed. Or is it?

Tri-State Crematory

Let's say you carefully lay out your cremation plans, include your family in the details, and pick out the perfect urn to hold your remains. Years down the road, your family visits the crematorium, knowing that everything has been taken care of for the family's peace. What if what you think happened, didn't actually happen?

Families that trusted <u>Tri-State Crematory</u> thought they were picking up loved one's ashes, not the remains of burned wood and concrete. Over 334 bodies destined for cremation at Tri-State had been dumped around the crematorium grounds and left to decay. The incident took place in 2002 in Georgia. When the bodies were found, many had decomposed beyond recognition and could not be identified.

Ray Marsh, the proprietor and operator of Tri-State, promptly received 787 criminal charges. The evidence was overwhelming, and he pleaded guilty. In 2004, Marsh was sentenced to two 12-year prison sentences from Georgia and Tennessee. The court saw fit to levy 75 years of probation after the sentences were served.

In a score of civil suits against the Marsh family and other funeral homes that shipped bodies to Tri-State, the families were awarded $80 million in the settlement. The Marsh's sold all of their property but still can't pay the tab. The collection of the $80 million remains to be seen. It never ceases to amaze me what people are capable of. But, what if it isn't people that shake up your eternity? What if it is a tsunami?

Tsunamis & Earthquakes in the Indian Ocean

In 2004, a massive earthquake in the Indian Ocean with a magnitude 9.3 started a chain fire event of several tsunamis stretching from Southeast Asia through the northwestern coast of Malaysia, coasting through the Indian subcontinent and reaching out to the African Great Lakes, killing nearly 300,000. The event was unprecedented, massive, and lethal.

The devastation and enormous body count resulted in thousands of bodies being cremated together to prevent disease, further adding to the chaos. Because of the disarray, bodies were not identified or viewed before cremation. Well, except for Westerners who perished. Westerners were kept separate from the Asian population.

To add to the situation's complexity, tourists from Japan and South Korea were mass-cremated instead of being separated and returned to their home countries for funeral rites. Mother nature is relentless and people are ill-equipped to deal with such a calamity. Without the historical significance of a natural disaster, a person's life and eternal resting place can be lost to history. It seems time heals all wounds - and buries the past.

Lost Cemeteries

History is a bit of violent business. There are wars, famine, scandal, disease, politics; you name it. The turmoil of history is perceived in varying degrees in different circles which leaves space for stories to get lost in the dark matter of time, including cemeteries. Not all of those 15 people underground for every 1 of us breathing have had their lives memorialized or commemorated. This absence is one of the most significant faults with how we do things today I am hoping to bring awareness to. Life marches on, and it marches right over eternal resting places, literally. Here are just a few <u>lost</u> & found.

Rome's Subway

Any time you kick a rock in Rome, you uncover a piece of history. This is the hazard of living in the cradle where civilization started. Imagine being tasked with digging a subway under the historical mecca of Rome! I would be tempted to dig with a spork for fear of damaging anything historical and being sued.

Rome found so many artifacts when digging their subway. These artifacts were large and small. On the larger side there was a 2nd-century home and a 2,3000-year-old aqueduct! The San Giovanni station opened a display showcasing some of these findings. The artifacts range from small ceramics to a 1st-century agricultural fountain.

Digging under Rome was hard work and presented several challenges when weaving around the dead underneath. I'm sure there is an extensive amount of paperwork that goes with that nightmare. While working on Line C, the subway crew found a 2nd-century military barrack with 39 rooms, complete with a mass grave of 13 skeletons. It is believed these were the barracks of the elite Praetorian Guard who protected the Roman Emperor Hadrian.

The London Tube

Digging subways in old countries leads to remarkable finds. For example, when the Crossrail commuter rail project was underway in London, the workers excavated medieval ice skates, a Tudor bowling ball, and of course, the mass graves created from the plagues that infested Europe. We know the mass graves were from plagues because the DNA in the skeleton's teeth still hold plague bacteria.

One of the mass graves found had 13 skeletons traced to the black death; the other had 42 traced to the Great Plague. Each person had their coffin within the mass grave to provide the dead a bit of privacy and dignity. These folks were just buried together in a quick pile out of fear.

Poltergeist Philly

By the name of this topic line, you probably had a vision run through your head; at least I did. And, you probably can guess what happened (insert line from Poltergeist movie):

"You son of a b*tch. You moved the cemetery, but you left the bodies, didn't you? You son of a b*tch, you left the bodies and you only moved the headstones. You only moved the headstones. Why? Why?" - Steve (Craig T. Nelson)

As construction was started on Arch Street in Philadelphia in 2017, coffins began popping up all over the place from a cemetery that was **supposed** to have been relocated. The First Baptist Church Burial Ground was established in 1707 and was believed to be relocated in 1859 to the Mount Moriah Cemetery. The *headstones* were indeed moved; the *bodies* stayed. More than 400 peoples' remains are currently being analyzed for identification.

Sweet Home Chicago

The Chicago Dunning area was infamous for its understaffed and overwhelmed institutions and insane asylums with patients receiving ill care; or no care at all and left to their own devices. The lack of oversight forced patients into abusive situations from staff and each other while living in squalor. Death did not require ceremony or a moment of silence. These institutions were aptly called *tombs for the living*.

The Cook County almshouse rested on 20 acres, complete with

a potter's field for the indigent and unclaimed. The soil embraced 100 unidentified deceased from the Great Fire of 1871 as well as society's forgotten. Over 38,000 bodies were uncovered in 1989 during the construction of luxury homes while laying sewer pipes. The bodies were relocated to Read-Dunning Memorial Park.

Florida - Just Wow

Tampa Bay has become the epicenter of raising the dead across the state in the last several years. In 2019, an African-American cemetery was uncovered, which led to a domino effect of lost cemeteries across Florida. The discovery was significant in Zion Cemetery and the Tampa Bay Times picked up the story making the public aware of the cover ups. More discoveries have blasted across the front pages of local media since that time.

North Florida Avenue contained 800 people buried in what is believed to be the first all-black cemetery established in 1901. But, unfortunately, Zion Cemetery had vanished in the 1920s under the development of Robles Park Village public housing.

The publication in papers created a spotlight, causing researchers to start investigating what happened and if there were more. Ridgewood Cemetery was discovered shortly after beneath what is now King High School in Tampa. Using ground-penetrating radar, over 145 pauper's graves were located beneath the school.

The Times kept rolling with the story and people started providing tips of cemeteries they remember being in certain areas that were now gone. That's when Clearwater Heights became suspect.

Clearwater Heights neighborhood and surrounding community residents reported they remember the graveyard being there. Residents also indicated there are unmarked graves in an empty field. Digging a bit deeper, College Hill Cemetery was found.

A map from the early 1900s shows an Italian Club Cemetery that is still open. The interesting thing is, to the left of where the Italian Club Cemetery is located, the map identifies a vacant lot labeled 'Colored Peoples' Cemetery'. The Colored Peoples' Cemetery was known as College Hill Cemetery. Next, the researchers located obituaries listing College Hill Cemetery as the final resting place of several people - over 100 obituaries were noted. And the dominoes continued to fall in Florida, leading to the finds of predominantly African-American lost cemeteries:

- Port Tampa Cemetery (MacDill Air Force Base)
- May-Stringer House Cemetery (Hernando Heritage Museum)
- Unnamed Cemetery in Clearwater on Holt Ave. & Engman St.

- Keystone Park Memorial Cemetery (Horse Ranch)
- Estuary Cemetery (Water Street)

Many projects are underway in Florida to identify remains and return them to the family for reburial. Or if no family can be found, to be relocated.

A Walk in the Park With COVID

We previously discussed how we are running out of room for underground burials. Recently, an article was published which raised the eyebrows of a few in recognition of this fact. According to <u>Motherboard</u>:

> *New York City is considering burying victims of Covid-19 in public parks, many of which are already built on top of burial grounds.*

Washington Square Park in New York City is a prime example. Below the sprawling green clipped grass and majestic landscape is a mass pauper's grave - population 20,000. These bodies accumulated between 1797 and 1825, with a majority succumbing to yellow fever epidemics. The irony is not lost on me that New York is considering temporarily opening the tunnel below the park to make room for bodies that fell victim to COVID-19.

The plan is to dig multiple trenches that would accommodate 10 caskets at a time in what will be called *NYC Park burials*. The Mayor has shied away from discussing the plans publicly because of the grim history, aka the mass pauper's grave we just discussed. To carry out these plans, the city is citing the 'land reuse' rule that we discussed earlier. New York and other cities have been developing over graves for years - as we have discovered, this is not new.

Elizabeth Meade, an archeologist, has spent decades documenting over 527 cemeteries misplaced throughout the five boroughs. Mead explained that Washington Square, Madison Square, and Bryant Parks are the best-known reclaimed areas to make parks. What do we mean by reclaimed? Reclaiming the cemetery title to repurpose the land for a park - or other city offering. Documentation from the 18th and 19th centuries support the findings. The city owns the cemeteries and has the authority to convert them to parks without much adieu.

Well, in one case, there was a little adieu. Central Park absorbed multiple cemeteries in its creation. A black community called Seneca Village was cleared and leveled to build the first public park in the US. Greenwich Village sits above St. John's Burial Ground. The city took both by eminent domain. The bodies *were not* moved. What is an eminent domain? It's flipping scary is what it is. The definition is:

> *"the right of a government or its agent to expropriate private property for public use, with payment of compensation."*

In well-established cities, it is common that parks or city land were at one point a cemetery. The phenomenon of repurposing land can be found from coast to coast. In 1957, San Diego passed a law allowing the government to remove tombstones in 'abandoned' cemeteries and convert them to public greenspace. There are similar laws across states.

Interestingly, Hart Island is occupied by over a million dead bodies and is under the control of the Department of Corrections. Why is this interesting? Because in December of 2019, the city council passed legislation that will put Hart Island in the control of the Department of Parks. It is not likely that the Department of Parks will continue to use the land as a prisoner burial ground. Cue the greenspace.

Parks are seen as a respectful commemoration of the dead. That's why I find it interesting that the living largely aren't aware of what lies beneath where they work and play. In order to recognize the former cemetery and to commemorate the dead buried there, the city will display a commemorative plate at the location. Usually the commemorative plate will be a bit vague; it's all in the wording. This is a bit eerie, but what really freaks me out is the things we do to preserve the dead. Today's preservation tactics tend to harm this, and future generations', quality of life. Vague commemorative plaques on a bench don't harm us.

CHAPTER 3

Toxins Last Longer Than An Eternity

As we continue our discussion, it is important to remember cemeteries have been built over by parks, schools, office buildings, and housing developments, among other things. This next bit is relevant to what we put in the ground, so we will ease into the topic with the earliest embalming method - mummification.

Mummification with the Egyptians

No discussion of burial would be complete without the ancient OGs of embalming and preservation, the Egyptians. Remember, the term 'burial' covers a wide range of possibilities. Bodies can be interred or placed in the ground in their final disposition in various containers or what I refer to as 'death Tupperware'. The where of the final resting place largely depends on the culture or religion.

Some cultures keep the dead in their backyard or an assigned close location so the dead can guide the living. Some cultures prefer to consecrate the ground. Still others believe in the old out of sight out of mind slogan and banish the dead to a discreet location. These locales are hidden from the world for several years just to be probed and scrutinized by archaeologists hundreds of years later. In some cultures, if you were considered evil or poor, you were in an anonymous grave. During war or plagues; mass graves were the answer.

Humans don't corner the market on burying their dead. Chimpanzees, elephants, and dogs have been observed doing the same. But, Egyptians were the first to practice embalming and mummification. In mummification, soft tissues and organs are preserved by nature or chemicals, to ward off decay. In the case of Egyptians, mummification was a deliberate act and practiced by several ancient cultures.

To prepare the body for mummification, all moisture from the body had to be sucked out. That's a big chore. The human body is 60% water up to:

- 73% brain
- 83% lungs

- 64% skin

- 79% muscles & kidneys

- 31% bones

Add a pouch of powder and we are jello. The Egyptians would suck out the water by coating the body in salt (inside and out). Wet things get mushy and mushy is bad if you are wanting to give your dead body a shelf life. I think this is why banana chips have no expiration date; unlike that gum you got in the old 1980 baseball card wax packs. I know some of you have stuck those in your mouth to relive nostalgia and paid for it (waits for nausea to pass).

Egyptians believed the body had to cross from the realm of the living to the realm of the dead in the most life-like resemblance as possible. If the dead body was not preserved, the person could not live comfortably in the afterlife. They needed all their body parts in working order. Considering we can view mummies that date back thousands of years, I would say they were successful. It is believed the mummification process was initiated around 2600 BCE and perfected over the course of 2,000 years. Not all mummies were created equal - there was a learning curve.

Making a mummy was no small task. Priests would treat, embalm, and wrap the bodies over the course of about 70 days, pausing for prayers and rituals. Removing the moisture included removing the internal organs. We saw how those internal organs are actually water bags. Priests were well versed in human anatomy. They knew how to use knives and hooked instruments.

The hooked instruments took special skill and finesse because they were inserted into the brain through the nose. The brain would be liquified, and pulled through the nostrils without disfiguring the face. Remember, the goal was a lifelike resemblance so special care was taken. That's not as disturbing as the fact that this process has been replicated at Universities - oh, joy.

Priests mastered making precise cuts in the abdomen so they could pull out all the other organs; except the heart. Egyptians believed the heart was the center of intelligence, not the brain. Organs were preserved separately in jars and buried with the mummy. As the craft of mummification evolved, the organs were preserved within the body after being removed and treated. These priests had surgical prowess and it makes me wonder if they used these skills on the living, but that's another book.

Once the body was dried out, it looked like a deflated football with things sunken in around where the organs used to be. No worries, the priests would inflate the body with linen and pop in a set of fake eyeballs. Perfect, a jerky looking shell of a human for afterlife consumption. All that was left to do was wrap it up in long strips of cloth. The wrapping was much like making a pinata.

Resin, wrap, repeat. Amulets, prayers, and scripts were written across the linen to protect the dead from misfortune.

Tombs in Egypt were usually started *before* a person died; talk about a perfect storm of micro-management in this ancient estate planning. Preparation for death started early in life because the Egyptians' had an overwhelming love *of life*. They lived in the present and saw their life as the most incredible gift. Egyptians wanted life to continue in the hereafter. This is a stark difference to what most believe as life being a mere stopping point to test us and determine our fate. The Egyptians believed life was the end all be all. Life was what's up. This made preparing for life after death of paramount importance.

If the tomb wasn't done at the time of death, construction would be ramped up to finish before the 70 day mummification period elapsed. Everything the deceased would need for the afterlife had to be hastily placed in the tomb. This was everything from furniture to wall paintings.

The Egyptians believed every item they put in the tomb would transfer to the afterlife with the deceased. This is like the deceased going to sleep in the tomb and waking up in the afterlife with all of their stuff intact and being able to interact with it. Once the tomb was packed like a UHaul cross country semi, final rituals would commence and the coffin would be placed in the chamber. Once everything was in place, the tomb would be sealed.

Egyptians went to great lengths to protect the body because the body was the permanent home of the soul. If the body was destroyed, all was lost. The Egyptians believed in three parts of one soul that lived in the body, similar to the Holy Trinity. One part of the soul would stay in the tomb, one could visit people outside the tomb, and one would cross to the afterlife. Even with these beliefs, not everyone was mummified. There were credentials to becoming a mummy, like *wealth*. People who were mummified were:

- Pharaohs
- Nobility
- Officials
- Select common people
- Sacred bulls
- Baboons
- Cats
- Birds
- Crocodiles

Mummification was expensive. Commoners that didn't have the funds would practice mummification methods. These DIY home remedies were considered poor methods and did not always yield desired results. We discussed the priests who had extensive surgical knowledge. Quality mummification takes sophisticated embalming skill sets.

Egyptian embalming required washing the inside of the body and the organs with precise concoctions of spices and palm wine. Although complex, mummy ingredients were generally not harmful to the environment because they were part of the environment and not synthetic. So, what happens when we use artificial preservatives? I know a lot of people that won't even eat artificial preservatives.

Short Story of Embalming - Long Effects

During the Civil War, bodies were piling up on porch steps and battlefields across America. The bodies had to be transported in mass quantities to hometowns for burial. Keep in mind, exposure to elements and time are not kind to a body and transportation is slow. To solve the problem, a commission was given to Dr. Thomas Holmes for his pioneering <u>efforts</u> in preserving bodies. Dr. Holmes was in the Army Medical Corps and embalmed over 4,000 soldiers in Washington, DC. His successes did not escape the interest of

President Abraham Lincoln, who was fascinated with the process of embalming.

Lincoln realized the impact of disease prevention and the immense commercial opportunity embalming could have, cha-ching. I mean, everyone dies, right? Shortly after Dr. Holmes' commission, embalming services were offered to the public. At the time:

1. Carpenters made coffins
2. Liverymen handled the dead; and
3. Embalmers embalmed

Being an all-encompassing Funeral Planner was not a thing. There was no cross-pollination between professions. Each profession kept to their own skill. That is until *undertakers* took up all three skills (carpentry, livery, and embalming) and funeral planning, including merchandise.

Since undertakers could embalm and incorporate all deathly business functions, people could plan for funerals without haste. Preservation allowed funerals to be more planned and meaningful without worrying about rapid decomposition or disease. During this time, embalming solutions were arsenic-based until formaldehyde was developed and available.

If you don't know, <u>arsenic</u> is a poison that when consumed in large amounts provides for a rapid death. When arsenic is consumed in small amounts, it causes serious illness or a prolonged death. The common denominator being DEATH. Which is OK if you are a dead person. But, if you are alive, let it be known the main cause of arsenic poisoning worldwide comes from drinking ground water containing high levels of the toxin.

Arsenic was great for delaying decomposition, but there was a cleaning agent being used in America that ended up being the preservation staple. Formaldehyde. Yes, people were cleaning with it. I think it was kind of like the duct tape of the world at the time. Duct tape fixes anything. In 1859, formaldehyde was known to clean anything.

There was only one tiny drawback, the exposure to the optimal killing organism makeup of formaldehyde was harming people when they were cleaning. As a result, a series of bans took place in the 1980s against the use of formaldehyde in products. Your hair would stand on end if we went into all of the products formaldehyde was used in. HINT: *everything*.

Anyway, with this dangerous exposure to toxins in formaldehyde, it was a good thing mobile embalming schools traveled the country in the 18 and 1900s. The embalming schools would provide 2 days of formaldehyde and embalming training to undertakers to get their certification. That's not disturbing, they are working on dead people. What is disturbing is embalming training time was equal to the amount of education for practicing doctors and dentists of the time. Trial through error is the best educator, no? Today you need eight years of experience and a degree to answer the phone. Crazy.

At the present time, embalming is seldom necessary unless the family wants an open casket. Besides preservation, if embalming is practiced, the main reason is disinfection to prevent dangerous organisms from escaping that survived the host's death. Embalming assists with restoration and a lifelike appearance. It is the lifelike appearance that eases closure in the family's psychology of death. But there is still a huge problem with preservation.

Although formaldehyde has been banned in product usage - it remains the Mrs. Congeniality in funeral homes, and funeral directors <u>prefer</u> it to new, less toxic methods available. Let's break down the risk. Formaldehyde is a carcinogen. According to Wikipedia:

> *A carcinogen is any substance, radionuclide, or radiation that promotes carcinogenesis, the formation of cancer. This may be due to the ability to damage the genome or to the disruption of cellular metabolic processes.*

That's right. It ranks right up there with tobacco, asbestos, radon, and other lethal items. ___Read that again.___ Then, think where most of these embalmed bodies go, not counting the ones under city park water fountains (we will be getting to that part soon, that section isn't for the faint of heart). All of these things are fine, if you are dead, but it is the harmful exposure to the living that is bothersome.

The reason for formaldehyde's popularity and high demand is it firms the body tissues more than any other embalming product out there. Funeral Directors are under immense pressure from clientele to have their loved ones look as life-like as possible. The death of a loved one is a tender topic requiring empathy, tact, and treading lightly on open wounds - emotions are high. Families want everlasting life from their funeral home, and formaldehyde delivers eternity. What are Funeral Directors to say at this most delicate time, No?

Undertakers are well aware of the risks of using the chemical and have taken necessary precautions to ensure their safety. Gloves, masks, eye protection, and ample ventilation are standard in the industry. Another consideration for its predominance is formaldehyde is cheaper and lasts longer than *greener* embalming solutions. The effects of formaldehyde is highlighted in studies of workers exposed to it having higher incidences of rare cancers.

There is <u>no</u> <u>dispute</u> among the medical community that formaldehyde causes <u>cancer</u>. For the medical community to agree so absolutely is unheard of. Yet, we see a massive amount of products we use, from plywood to skincare products with formaldehyde in it. The use has gone down, but we still wonder why cancer is so prevalent in society and examine everything but the dead for the answer. Using toxic chemicals like this is a far cry from where we used to be and cancer was around, but not as prevalent as it is. We seem to create our own drama.

Back in the day, family and friends would wash and dress the body for viewing and paying respects in the parlor or back yard. It was an intimate, personal affair. The body was buried before the natural decomposition started. Burying the body was seen as returning the body to the earth, to <u>nature</u>. In no way am I up for my children seeing me naked and preparing my body, that isn't what I am saying. There would not be enough funds in my estate to cover the psychological damage to my children.

What I am saying is, back then there were no lethal chemicals bloating up the deceased. People were not afraid to eat the food grown around the family gravesite or drink the water from the well that was downhill from a graveyard. The very commercial and expensive process we have today is full of toxins, not only chemical toxins. The death business is full of high stress and fatigue for the family and plagued with the uncertainty of whether the body will rest in peace. That is a lot of downside stemming from a $15 Billion funeral industry. Not to mention, a lot of physical harm to the living.

What Are You Drinking?

Caskets composed of wood, metal, and luxurious satin pillows weigh in at a few hundred pounds. Generally, the more celebrated a person is translates into how much their casket weighs.

For example, Ronald Reagan was buried in a mahogany fortress of over 400 pounds. Some caskets are encased in 3,000 pounds of cement or steel. The amount of material weaved into the death business every year could build 4.5 million homes and is equivalent to 4 million acres of forest. This is a gross misuse of valuable resources for the *living*.

That isn't all, added to the weight of the body is toxic embalming fluids. Fun fact, some embalming fluids are hallucinogenic which do not alter state and stay in the earth forever. These fluids average one gallon for every 50 pounds the human weighed. This calculation brings the total weight of ONE body to about 2 tons. Nearly 800,000 gallons (more than an Olympic-sized swimming pool) of formaldehyde are soaking into the ground with bodies every year.

Although embalming liquids slow the process, bodies still decompose. It doesn't matter how much wood, metal, or concrete they are encased in, decomposition happens. Leaks happen as materials break down. Where does the decomposition sludge of toxic substances and bodies go when they are in the soil? Our water table. Fluoride is the least of our worries. All kinds of resources and pollutants are tied up in death.

Memorial Parks and cemeteries require tremendous upkeep, chemical fertilizers, pesticides, and water to maintain the pristine features. Remember, it is most likely a Park now, and toxins from the dead and the maintenance are lurking below. If the chemicals below don't kill you, the chemicals above will. There is no need for the funeral industry to secure their jobs in this manner - death is a certainty.

Cremation is not off the hook in taking a toll on the environment or poisoning us with toxins. Cremations have gained popularity and are fast becoming the preferred and affordable remedy compared to a $15,000 - $25,000 casket and added funeral expenses. The Federation of British Cremation Authorities (yea, I didn't know that was a thing either) reported cremators function on 760-1150C for 75 minutes per cremation using 285 kiloWatt hours of gas and 15-kilowatt-hours of electricity per cremation.

That's a metric mouthful, let me 'Merica that up. The energy and resources used during *75 minutes* of cremating a person is equivalent to the energy and resources one person uses in a *month*. It's not only energy used. Cremation is responsible for 16% of the UK's mercury pollution. Mercury? Yup, dental fillings.

Bodies are pumped full of formaldehyde and cremated inside a coffin. The burning fumes make their way into the atmosphere instead of the soil and water table. It seems, either way, we are negatively impacting our environment and ingesting toxins. FYI, ashes from cremation have zero benefits for putting nutrients back into the earth if you bury them. Cremation burned any available nutrients to soil.

It has been discovered that our past does come back to haunt us. Rotting corpses from the Civil War era are becoming a big problem for homeowners. Be warned that toxins can leak out of the grave, pollute the drinking water, and cause serious health issues. You, your family, and your neighbors may be downing a non-degradable arsenic and formaldehyde cocktail without knowing it. Bodies degrade; chemicals do not. Arsenic seeps into the soil after its host dissolves.

> *"A Civil War-era cemetery filled with plenty of graves—things seldom stay where you want them to,"* says Benjamin Bostick, a geochemist at Columbia University. *"As the body is becoming soil, arsenic is being added to the soil."* From there, rainwater and flooding can wash arsenic into the water table.

The federal government documents that it is safe to drink water with 10 parts per billion arsenic levels. Iowa City exceeded that amount by three times in 2002 near one of the old cemeteries. Formaldehyde (a carcinogen) replaced arsenic (a carcinogen). Test your water, folks - be Erin Brockovich. According to <u>VICE</u>:

> *David Spiers, founder of Greenacre Innovations, a company specializing in burial technologies that protect the environment and people, has been trying to raise the alarm on formaldehyde leaching and the dangers it poses.*
>
> *He made it clear that this was not just a Northern Irish problem but a worldwide one: "There is a much larger threat posed from all cemetery leachate including the formaldehyde element, regardless of where these are situated in the world. Not enough investigation*

has been carried out in this sensitive area. Contrary to being told it dissipates over time, [formaldehyde] actually merely dilutes, leaving the highly probable conclusion that some percentage of this carcinogen toxin may well make its way into some of our ground water source."

It may be time to rethink what we do with a dead body.

CHAPTER 4

Things You Can Do With A Dead Body

There are a lot of old sayings and morals of the stories that point to leaving things better than we found them for future generations. It is something that is taught in Kindergarten and is right up there with 'Don't Bite Your Friends'. When I was younger and someone was giving me a barrel of righteousness, I would shrug my shoulders and bob my head in agreement, not really understanding the depth of the proverb or what it was saying. As time passed, one of my favorites became:

A society grows great when old men plant trees whose shade they know they shall never sit in
---Greek Proverb

It wasn't until I had kids of my own that I realized I wanted better for them. It was when I was standing in front of a fireplace with my son discussing our dog's ashes. It killed me to see the absolute disappointment on his face and enlightenment jolted me into wanting better for my kids. I understood the depth of benefits in planting a tree I would never see.

Life wasn't all about the material things I provided them at the moment. It was in the wisdom of my actions that would guide them long after I passed. My son let me know that I had been entrusted with a part to play in society to shape the world he and his children would grow up in. The realization was profound. I am not that person, but looking at him, I knew I had to be.

I was going through an identity crisis. It is not the first; or last time to be sure, that my son initiates and demands my personal growth. I am not a vegetarian, or wear hemp clothing, or live minimally - but I found myself legitimately concerned about parks, drinking water, the environment, and eternally resting in peace. The more I thought, the more I realized being in an urn was not for me. Something I had not thought about as very pressing suddenly became a priority to wrap my head around and give meaningful attention to.

I had lectured my son on more than one occasion to pick up after himself, help clean up at a friend's house, turn the lights off, don't leave the water running, look out for everyone smaller than you -- the overwhelming feeling that I don't do what I ask of him was abrupt and sobering. So, what was the solution? You can't fix something that doesn't have a shelf life. We still have a substantial nuclear radiation problem we can't seem to tie off - but that is another book. For now, we can start with the little stuff - dead bodies and toxins.

Green Funerals

People are starting to recognize the stress placed on the environment from traditional burials and are opting for green burials with biodegradable coffins and headstones in the form of rocks or trees. The toxic embalming fluid is replaced with a plant-based concoction. The appeal is not only sustainability but also saving the land for the living.

The Green Burial Council provides a <u>list</u> of State locations participating in green burials if you are interested. Green burials are designed to leave the land looking untouched and not like a cemetery while protecting its resources. The locations are quite lovely and don't look like a cemetery. Location aesthetics are peaceful and park-like. How ironic, right?

Memorializing the dead is becoming less traditional and more ceremoniously meaningful. This is especially true of today's generations who are more in tune with collective planetary restoration and sustainable solutions for dwindling resources.

Burial Belts

Australia and a few other countries, including the US, have talked about implementing burial belts on the outskirts of cities. The theory of burial belts is if bodies are buried without harmful chemicals, they assist green space and vegetable gardens to be planted alongside them.

Animal lovers are particularly fond of this idea because it protects the wildlife living there, ensuring a habitat without urban encroachment. Burial belts hit three heads of the demon: Toxins, multipurpose space, and wildlife preserve.

Death as a conservation effort is a practical idea. I tried to think of an argument against it and really couldn't without sounding like an asshole and making up reasons. Taking care of the ecosystem, protecting human health, and ceremoniously returning to the land - it feels wholesome, right. I bet this is how Superman feels. South Carolina and Texas state departments have already started these conservation-type burials to save the land from development and enveloped burial belts into the park system.

Blossom Into A Tree

The thought of being a tree makes me smile. It could be the Game of Thrones nerd in me with the whole 'old gods' thing, but being a tree appeals to my romantic nature. I love the thought of being memorialized in a cherry blossom tree. I mean, how beautiful to celebrate my life, right?!

Bio Urns are a biodegradable urn that turns you into a tree.

The urn has two capsules, one for the seedling and the other for your remains that the seed will root into. You get to be anything you can grow. That is the cool thing about <u>Bio Urns</u>. Bio Urns allow you to:

- Use any seedling of your choice, even plants or flowers, if trees aren't your thing
- Make a positive impact on the environment
- Be placed in parks dedicated to using Bio Urns
- Commemorate loved ones in a meaningful way
- Use Bio-Urns for both humans and pets
- Participate in an economical and easy to do process
- Use any cremated remains (including the vase on the fireplace)

Bio Urns give you the capability to plant the seedling immediately, or wait until the seedling is better established, or keep the plant to keep your loved one close by using an incube lite. Bio Urn offers eclectic solutions for minimalist lifestyles or people without a backyard.

Choosing the seedling you want to be is essential not only for sentimental reasons, but you want to make sure your family and friends won't require a horticulturist degree for your maintenance and care. In addition, choosing something indigenous to where you intend to root is vital in observing ecological balance.

Bio Urn's focus has been *to add meaning to this process of life and return to nature*. The standard urns are 100% recycled materials sourced locally. Over 47 countries in 5 continents have embraced the idea of celebrating their loved ones through rebirth in trees, plants, bushes, flowers, and many other types of seedlings and sprouts.

Humans and pets fused with nature promote reconciliation with the planet that intimately provides a prosperous life (romantic, le sigh). Seeing myself in this cherry blossom tree makes me believe the grieving process will be easier for my children because I continue to live. The thought of my demise is far less depressing, even for my soulful middle child.

Bio urns work by allowing the seed to germinate separate from the ashes until the plant or tree roots are strong enough to interact with the ashes. This is measured and timed with the decomposition process of the urn. The urn planters don't have an expiration date, so you can purchase them whenever you like.

There is a downside, it requires cremation which means burning toxins in the atmosphere. Unfortunately, several of these alternatives require cremation. I also imagine this is how haunted forests start. The good news is there is an alternative way to become a tree without cremation. However, you will need to check the rules in your state before committing to a body pod.

Body Pods

<u>Capsula Mundi</u> incorporates the ritual of whole-body transformation versus ashes. That means no cremation. An egg-shaped pod made of biodegradable material encapsulates the loved one's body positioned in the fetal position. The egg is ceremonious with beginning and ending the same way we start, we finish. The egg is seen as the perfect ancient capsule.

I know what you are thinking; *I've seen this movie before* as remnants of *Soylent Green* play in your head. Don't worry, the process is much more comforting. How it works is the loved one chooses a tree before their death. The tree will be planted on top of the egg serving as the headstone for the deceased.

Using a body pod transforms the way cemeteries look today into a warm, receptive celebration of life while preserving valuable resources and caring for our planet. Capsula Mundi's focus is to reconnect society with nature and redefine death from a taboo topic into a reverent celebration of life.

This alternative is much like using biodegradable caskets and participating in green burials. The result of Capsula Mundi is a coffin using ecological materials and significant symbols such as the egg and the tree to reconnect us to nature. The tree is renowned for connecting the sky and earth. The process to commemorate a loved one is relatively simple and budget-friendly.

Capsula Mundi offers urns for ashes as well. If the name didn't give it away, the company is Italian, and the body pods are still in start-up phases but gaining momentum quickly. At least on that side of the world. These burials are legal in many countries, but not everywhere. As green cemeteries take over, Capsula Mundi is hoping to raise awareness. I'm rooting for them!

Diamonds Are Forever

Creating a diamond from your ashes to pass down as an heirloom is an elegant way to be immortal. Eterneva makes this dream a reality and takes the family through a complete journey from ashes to diamond. Eterneva is invested and fully immersed in the deceased's life story.

Upon payment of $100 for the welcome package, your journey begins. Eterneva has an inauguration for your loved one. The inauguration entails adding the picture of your loved one to Eterneva's wall and sharing the story of their life with the Eterneva team. Eterneva also memorializes the loved one with a dedication page on their website to remain for all time.

The family is walked through the entire process with videos starting at the inauguration. The next video is about the ashes being purified into carbon, followed by the growth of the loved one's carbon into a diamond. Updates are given throughout the process, and the final video is the homecoming. Eterneva hand-delivers the diamond.

During the process, you choose a diamond color that means something to you. It can be the color of the loved one's hair, eyes, or their favorite color. Any color that means something to you. Eterneva can engrave the name on the side of the diamond for deeper personalization. The process is a bit more expensive than a traditional funeral cost, but you get so much more out of it. The value of this experience is hard to put a price on.

Diamonds are a symbol of eternity, which is why tradition dictates to give them as wedding rings. The unbreakable stone is a symbol of unwavering commitment in love. What better way to keep a loved one with you and pay tribute to their memory? Diamonds are a physical representation of the spirit, affection, and charisma of the deceased that touched your life.

Oceans Eternal: Be A Reef

If you love everything about the ocean or are an avid Titanic fan, Eternal Reefs might be the way to go in your immortality. Reef balls replicate Mother Nature in the ocean world by encouraging reef development and supporting ocean life. They are active in more than 70 countries to the tune of 750,000 reef balls in the oceans.

Reef balls invite microorganisms to land, burrow, and mature, which eventually propagates in sufficient numbers for predators to feed on them. Fish immediately occupy reef balls when they are dropped and find their forever home. This forever home is ideal for a permanent living legacy to commemorate the ashes of a loved one.

Creating a memorial reef serves as a medium that the loved one can contribute to the marine environment for eternity. During the casting, family members place handprints or memorabilia in the soft mixture. Military burials are also popular in this fashion.

Reef balls mix the ashes with an environmentally safe cement mixture explicitly designed for ocean habitat to create artificial reef formations. Families select from one of the permitted locations to deploy the reef ball and choose from one of the many take-home mementos they offer.

Eternal Reefs is focused on creating a positive experience to heal the mind, sea, and soul in preserving, protecting, and enhancing ocean life. There are several ceremonies during the process to help in that endeavor.

A pre-cast reef ball is prepared and presented to the family for mixing remains and creating a 'pearl,' the centerpiece that fits inside the reef ball. After the reef ball is cast and finalized, a viewing is held where:

- Pictures are taken
- Rubbings of the bronze plaque are encouraged to take home
- Final goodbyes are given; and
- Tributes are written

A placement and dedication ceremony is held where a boat takes the family from the reef site to where the ball will be deployed. The family is given tribute reefs (miniature reef balls) as a devotion to their loved one and provided the GPS location of the drop.

Eternal Reefs are an excellent alternative for pets; well, probably not the family goldfish due to the costs, which aren't outlandish (see what I did there) but do have a comma in it.

For this reason, Eternal Reefs encourages holding your pet's remains and ensuring the pet and family members are put together in a reef at a later date with no additional cost.

There are a few other ways to commemorate your life but they are not as permanent and sustainable as the ones just discussed. However, they are pretty righteous:

The Haunting

If you want to be a creeper and haunt your family, you can have your ashes compacted into a vinyl record for about $3,800. The record will play 24 minutes of music or an audio recording (http://www.andvinyly.com/)

Light It Up

The UK is creative in ways to go out with a bang. Your ashes can be scattered in the sky courtesy of fireworks by Heavenly Stars Fireworks. They offer a complete Remembrance Display package for about $600 that explodes into dragon eggs, flowers, and peonies. They have various pricier packages if this isn't enough of a send-off and you need more thunderous applause.

Be Artsy

Thanks to Art From Ashes, you can become friendship balls, suncatchers, paperweights, glass ornaments, and much more. Ashes are infused into a specialized glass blowing technique. It just takes a teaspoon of ash, so technically, you could create a chess set or your favorite football teams' figurines. The name of your loved one can be etched into sculptures. Most of these ash arts you would never know incorporated ashes unless someone told you.

I Need A Hero

Cremation Solutions does it *ALL*, but what captured my heart and sent my inner geek into euphoria were the 12" action figures in your likeness! Fandom reaches out from death and seizes the opportunity for your ashes to rest in Superman, Indiana Jones, and more for about $250. If you want something sophisticated, Cremation Solutions creates custom portraits in your likeness by adding a small amount of your ashes to the painting.

One of These Days Alice

I'm not saying hurry up and die, but you can register for the next launch date to outer space. Celestis provides the unique opportunity to honor loved ones with exploration, adventure, passion, and the cosmic universe.

Celestis not only launches ashes into space but a memorial spaceflight is not limited to those choosing cremation; any final disposition is welcome. Additionally, space flights are available for off-planet DNA storage packages.

Packages start at $1,295 for a portion of your ashes to be launched into space and returned to earth or $4,995 to be launched into Earth's orbit. If you prefer to be in the moon's orbit or on its surface, the price is $12,500. Loved ones attend the launch and are provided a tracking tool to watch your spacely progress.

Do Like Han Solo

Alcor and its members believe death is not an instant permanent state. Death is permanent when memory and personality are disconnected, and the person cannot be recovered. They believe that transitioning *living* to *death* is not instantaneous. They might not be wrong. Brain activity has been documented to persist after death is declared. Theoretically, your brain knows you are dead - which means your conscious mind has that knowledge. That is scary af.

When death occurs is contested. One study in 2011 involved decapitated rats. Electrodes attached to the rats heads showed consciousness and sensory awareness observed in all rats post-decapitation for up to a full minute. The lag leaves room for cryonics. It sounds like something right out of *Flatliners*. In Flatliners, medical students tried an experiment to see if there was life after death. Just as in the rat study, the brain continued firing and the students brought back near death experiences and the consequences of playing with fire.

Alcor believes the brain still fires. If you become a member of Alcor, they will ensure that professionals are at your bedside to start the process as soon as you are declared dead. What process? Cryonics.

Cryonics preserves life by pausing death through freezing. The theory is that once a cure or medical technology advances for what killed you, a dead person will be restored to life to receive the treatment. Alcor is considered an *ambulance to the future*. Like most ambulance rides, this one is insanely expensive.

The membership dues aren't bad at $50 for you and $27 for each additional family member per month. Cryonic costs used to be paid by living relatives, but that process was determined unsustainable, so now there is insurance. Over 90% of Alcor members have life insurance policies naming Alcor as beneficiary.

The requirement is a minimum of $200,000 for the whole body and $80,000 for neuro (brain) life insurance. Policies that increase with inflation are a wise decision and encouraged as future costs are unknown. Alcor recommends insurance reps that are familiar with cryonic costs.

Upon death, Alcor will transfer significant funding from the patient to the Alcor Patient Care Trust to pay for storage costs. Alcor then moves investment funds into separate Alcor Care Trust Supporting Organizations held at Morgan Stanley.

The accounting is designed for patient storage costs to be taken solely from the income of the Trusts to ensure funds for indefinite storage. The whole thing is hella cool, and if I had the money, I would.

Cloning

Sorry guys, this one is only for animals. But still, who knows, am I right? Barbara Streisand gave a <u>Variety</u> interview in which she revealed she had two of her dogs cloned after one passed. There is a wide range of people for and against the process of cloning.

The procedure clocks in around $50,000 for dogs and $25,000 for cats, more for horses. How cloning works is the DNA is extracted from the pet and cryogenically preserved until a host, or I should say surrogate, is located. Then, the DNA from the surrogate's eggs is replaced with the preserved pet's DNA which alters the eggs.

If all goes well, the surrogate becomes pregnant and carries the clone to term. The surrogate goes through several tests, multiple pregnancies, and chemical alterations to make an ideal climate for the clone.

The process of cloning is an uncomfortable existence for a surrogate and one of the main reasons in the against column. The hormonal supplements given allow the surrogate to create embryos at will. Cloning doesn't guarantee the same personality as your pet's that you fell in love with. You could end up with an animal with a completely different temperament.

One of the companies providing cloning services is <u>Sinogene</u>. They proclaim to be a worldwide animal cloning expert and have successfully cloned hundreds of animals. Sinogene hopes to expand breeding and genetic technologies. It's a slippery slope, but if you want to ride it, cloning is a possibility.

Donating Your Body To Science

Most people have involuntary spasms when they think about what happens to your body when you donate it to science. A gory scene unfolds in your mind that makes Faces of Death look like a lullaby. Let's massage that horrid vision out of your head regarding contributing your body to the advancement of science. Let's start by answering: *Why Do It?*

First, you can <u>donate</u> your body in several ways: body, organ, and tissue. When you donate your body to science, researchers better understand the progression of disease and how to treat or cure it. Donations have lead to advancements and genuine breakthroughs in things like:

- Alzheimer's
- Cancer
- Heart Disease
- Diabetes
- Parkinson's disease;
- And More!

Donations help future doctors master their craft and benefit the living with their expertise. You can donate to a specific University or Medical center by contacting them and asking for a registration package that details everything involved. If you are unsure, the <u>state anatomical boards</u> can help you locate a University and body donation program that conducts research that aligns with your interests.

NOTE: Being an organ donor designated by the checkbox on your license *is not* the same as donating your body to science.

In some instances, you can do both. In others, your body may be rejected from donation. Having a Plan B is crucial if you have not left clear, concise agreements and instructions with a body donation organization.

Donating your body requires far more paperwork and permission than checking a box. Both organ and body donation are time-sensitive and must be planned in advance. Planning means informing and including loved ones in your decision-making and wishes. Helpful resources from accredited facilities would be:

- Anatomy Gifts Registry
- United Tissue Network
- Research for Life
- Science Care
- Medcure

As with any company, *do your research,* so you know what you are getting into. If you do not want your body pieced out (yes, that happens-specifically in for-profit companies), make sure you address that in the contract. There are laws to prevent selling human body parts by the piece; but there are always loopholes. There are no givebacks. Once your body is donated, if your family follows through with your wishes, any authority on what happens to your body is lost. There is not a lot of oversight in the middle-man body donation industry. The one thing you can be sure of is someone will benefit from the knowledge gained, whether your body is in pieces or intact.

Donating bodies isn't for everyone. And not everyone can donate a body. Some immediate conditions will make your body <u>unfit</u> for donation to science (again, Plan B). You will want to check with the organization to be certain. For example, dieing of a communicable illness will most likely vote you off the donation island.

Remember, a donation is time-sensitive. The first thing doctors do is test for infectious diseases before popping you in the freezer. Some organizations will embalm; others will not. This depends on the type of research being conducted. If you died in a horrific accident that rendered organs useless or damaged parts of your body, you might be rejected.

Researchers seek bodies resembling the average, healthy adult. I know what you are thinking because I thought it too. If I was in perfect health with nothing destroying my organs, why would I be dead? It can be a puzzle, and the answer is above my philosophical paygrade.

There are exceptions that will be detailed out in any donor form or contract. However, if you die under suspicious circumstances and an autopsy must be done - that will most likely disqualify you too. Science has recently started calling for more diverse body types (obesity may cause a rejection) so students and doctors can better understand how obesity affects disease and vice versa.

If all is well and your body is inducted into the scientific community as a bonafide member, the organization will usually take care of all the associated costs. If not, the costs will fall to your estate or family - so that Plan B is no joke. The family must approve the arrangements regardless of what you preordained. Therefore, discussing your plans with them is essential. According to the Forensic Anthropology Center of the University of Tennessee:

> *Regardless of what you have arranged, signed or instructed, your family or next-of-kin has the final say. We will not fight your family for your body. We urge you to convince your family that the donation is what you want at your death.*

Opting to donate your body to science has the potential to help millions of people through research, study, and implementation of a cure. This is the most considerable fact that separates organ donation and body donation. Organ donation is life-saving, whereas body donation is life-giving. Your contribution makes a difference, whether it is a safety test cadaver for car companies, a skeleton for educational institutions, or contributing to a cancer research lab.

CHAPTER 5

Losing Our Past & Who We Are

Do you remember Blockbuster stores? When digital venues came onto the scene, like Netflix and Amazon, I knew Blockbuster was in its final days unless they stayed ahead of changes. Blockbuster did not see their demise as clearly. Some scholars are comparing alternative methods of funerals to be the Blockbuster in the green burial equation. There is a wide belief we will lose who we are.

I think the loss of who we are has already occurred on a large scale. Entire cemeteries are LOST. History, knowledge, and stories are LOST. We are blessed with the many technological advancements that could serve as a viable solution to lost information. The ability to preserve our history comforts me (we will get into the how later).

What I struggle with is the *future* we are losing because of the current status quo. We have established how the dead are invading and poisoning the living. Why can't we have both historic preservation and a future of commemorating our loved ones without killing ourselves or the planet?

I don't think the scholars need to turn over in their grave just yet. Only 5% of today's burials are green; that's a great start, but not enough. What is promising is that 72% of cemeteries report an upswing in demand for alternative practices. Nearly 54% of Americans said they would consider green burial options. It's time to push the envelope and bring awareness. Things put into the ground should decompose without killing us.

In a lot of ways, we would not be lost. Humans would reach for digital immortality, not toxic immortality. Digital footprints last forever, right? It is practical, generationally accountable, and financially responsible to opt for alternative methods. We take personal accountability by preparing our future death and not leaving it to family members. The focus should be on the celebration of life. Don't just say the words; complete the actions. Be remembered well.

In the scholar mindset, it is believed that disposing of cemeteries will take away from the emotional depth of a city as cemeteries are spiritual places. I don't wholly disagree with this; places of death receive significant funding every year through tourism and donations. Let's go through some of them.

The French Catacombs

Romans have stunning catacombs. Rome's catacombs were intended to house the dead as part of their culture after burning bodies went out of style. But, it was Napoleon who had the foresight to design the French Catacombs into a tourist attraction. It didn't start that way.

To set the scene, <u>Paris</u> snowballed into the epicenter of art, life, and everything worldly. A significant number of growing pains go hand in hand with a tremendous amount of growth and death, especially when disease infects a population faster than the cure. The cemeteries were overflowing, and in some regions of the city, streets reeked of decomposing flesh. You would never have thought that luxurious perfume came from the same country.

To exacerbate the problem, the streets most affected by death surrounded marketplaces. Marketplaces where people bought, sold, and prepared food. The cross-contamination grew to epic proportions. Often, the craziest idea is the best solution, and that is what happened here. In the act of architectural magnificence and mind-numbing grossness with a flair for the macabre, Paris was bravely transformed into a mecca of progress from the cesspool of decay and despair it had become.

At the time, approximately 6 million bodies were pulsing under French cemeteries. The Cemetery of Innocents was holding 2 million of them alone. The idea was to handle these bodies with respect and a sizable ritualistic acknowledgment. The solution came down to the dead being relocated for the greater good of the public. As a result, the Paris Catacombs are historically one of the most extensive transmissions of human remains in history.

Paris sat on a vast network of 200 miles of limestone tunnels. The tunnels were previously carved out to obtain the stones that built Paris. The solution is beautiful equity in a way. The underground tunnels gave life to the city above *and* opened a space for the dead to remain part of it.

The stones that built Paris were weighing heavily on the ground right above the tunnels. These stones were so heavy that the threat of the stones collapsing through the earth and into the tunnels was imminent. The tunnels were extensive throughout the city and the buildings weighing down on them lead to giant sinkholes appearing across the landscape. Buildings were collapsing. This was a huge problem.

Charles-Axel Guillaumot was given the task of stabilizing the system by the King in 1777. His job was to prevent the collapse of Paris into the tunnels. Generally speaking, people who disappointed royalty never fared well, and saving Paris was no small task. A lesser man would have panicked - well, me. I would have panicked. But Guillaumot looked at the whole of the problems facing the city to see if he could solve two problems with one stone and earn the favor of the King. He pulled it off like a boss.

Fast forward to 10 years later, Guillaumot had saved Paris *twice* over. His plan was to schedule a nightly adventure to fill the tunnels and support the weight while solving the fleshy smell throughout Paris. I guess that makes it worth saving. Every night, Guillaumot ensured a priest chaperoned wagon loads of remains being ported from cemeteries to the tunnels. The priest chanted prayers while bodies traveled to the tunnels.

The nightly shift lasted for 2 years for the transport of remains and until 1859 for the transport of bones. This type of action was unheard of at the time and to keep it on the downlow and minimize public scrutiny he kept the shifts small, respectful and under the cover of darkness. Guillaumot prevented anyone digging while he swooped in with his cape doing the top secret dirty work, and promptly placed public utilities on hold during the French Revolution.

A small change occurred during this time, Napoleon was now

in charge. Being an ambitious sort, he rode the wave of the city's rapid rebuild and modernization - now that it wouldn't sink. While sitting on the throne as Emperor, Napoleon made one of his core beliefs known:

> *"Men are only great through the monuments they leave behind them."*

Maybe you have heard of the Napoleon Complex? Napoleon was height-challenged and had a scorching, fast temper. The complex is a belief that short men compensate for the lack of height with grand gestures laced in domineering, aggressive behavior. Rome had a sophisticated catacomb system. Rome also had an enormous vault of monuments and statues.

Napoleon wanted to incorporate both of these features in the newly acquired Paris Catacombs. He wanted to showcase the greatness of Paris for all the world to see. The catacombs were now part of a dick measuring contest. The goal was to attract tourists in droves and the idea was to manipulate bones into grand structures and art that was unique then-and now.

Although transferring 6 million bodies into the tunnels occurred with a ceremonial, organized procession, the bones were not placed in the same fashion. Bones were scattered in disorganized heaps throughout the tunnels. My OCD radar is on high alert with this information.

The quarry-men Napoleon hired had to sort the bones and artfully assemble everything they found in intrinsic, meaningful, and grand designs. In an effort to assist tourists, signs were carved by candlelight into the ceiling so people would know the way to go. Remember, the tunnels almost sank Paris and they are vast. It is easy to get turned around. For example, despite the signs, in 1903, 25 English tourists got lost, but there's always one in the group, right?

The official brochure to tour the catacombs was published in 1810 and swiftly gained popularity among the tourist community. Humans are a macabre group. It wasn't long before illegal parties were held in the catacombs. Full-blown raves were held as early as 1897. Where there are parties, there is mischief. Skulls have been stolen from the catacombs. The maintenance workers grew tired of replacing the stolen skulls so now you can identify where skulls were stolen by the gap in the design found in spaces throughout the tunnels.

Napoleon was a fan of the elaborate designs and incorporated exhibitions in the catacombs to draw even more tourists. Exhibitions include a deformed skeletons room, rare mineral displays, stone carvings, and fountains, among other things.

Mutter Museum

Nestled in the brotherly loving arms of Philadelphia, the Mutter Museum pays homage to the human body and the painful challenges that took place in history to understand it. The walls of the museum boast a rich history speaking to the medical profession's trials and tribulations of diagnosing disease and pioneering efforts in the advancements of medicine.

Mutter Museum offers fascinating permanent exhibits, as well as, entertains temporary exhibitions. I was delighted to see Mutter create virtual tours, which can be viewed under the website's *Education* tab. Virtual tour options include:

- Tour of the permanent exhibit
- Spit Spreads Death: The Influenza Pandemic of 1918-19 in Philadelphia
- Imperfecta, examines the shifting perceptions about abnormal human development, from fear and wonder to curiosity and clinical science
- Broken Bodies, Suffering Spirits: Injury, Death, and Healing in Civil War Philadelphia

Mutter Museum also features several online exhibits such as:
- Memento mori– "remember that you shall die."

- Healing Energy: Radium in America
- Under the Influence of the Heavens: Astrology in Medicine in the 15th and 16th Centuries
- History of Vaccines

The Mutter Museum prides itself on keeping the public disturbingly informed. I would have to agree as their last Podcast delved into the wild world of scrotums (with bonus scrotum trauma-don't google that!), forensic toxicology, corpse catapults, and how to donate your body to science. It seriously had me on the edge of my seat with a few grimaces here and there.

Body Worlds Exhibit

I had the eye-opening pleasure of viewing this exhibit when it stopped in Las Vegas. I wasn't sure what to expect when I handed over my ticket and walked inside. I was stunned, intrigued, disturbed, and empowered to change some of my more destructive behaviors. The exhibit smacks you with a massive dose of reality. It is an unforgettable experience.

Body Worlds peels away the exterior of humans that we see every day and gives an intimate look at the inside. The interweavings of organs and systems are on display which helps you identify with your own body. A side-effect of seeing the exhibit is an awakening. In their words:

The exhibition aims to inspire visitors to live consciously, to pay more attention to their health, to recognize their physical potential and limitations, and to reflect on what it means to be human.

Truer words have never been spoken. What you see are actual humans that donated their bodies to science so we could learn and be inspired from them. The exhibit navigates organ functions, diseases, and consequences of lifestyles that are not healthy. Body World explains the science and correlations in a way that is easily understood and digested by the viewer.

Commemoration with Celebration

Many cultures dually keep their identity and remember their ancestors in celebrations every year. Once I started thinking about it, I was amazed - and a bit miffed - that America has not embraced the tradition of an Ancestral Celebration Day.

Americans should stop cringing at history and start celebrating what people, all people, have contributed to this world as a testimony of their existence. Cultures that celebrate their past recognize the value of knowing and honoring those that lived before us.

The concern for losing ourselves is misplaced as we started that process a long time ago through the extermination of history that blotches or sheds a poor light on our nation. Some cultures know how to pay proper respects:

- The <u>Chuseok</u> in South Korea is a day to give thanks to ancestors for a good harvest.

- All Saint's Day is celebrated after Halloween by Christians recognizes patron saints and past loved ones. In Latin America, this day is El Dia de Los Muertos, where the celebration of life can be viewed in several cities drenched in colorful costumes and displays.

- Gaijatra in Nepal is an eight-day feast to commemorate the deaths of people that occurred during the prior year.

- Ari Muyang in Malaysia is a day of costumes and thanking ancestors for good fortunes and prosperity.

- Bon Festival in Japan commemorates ancestors with frivolity, fireworks, games, and dances.

- Pitru Paksha is a Hindu celebration lasting 15 days with food offerings and remembrance of ancestors and loved ones.

- The Hungry Ghost Festival in China is a MONTH-long celebration. Feasts are ample, and empty chairs are often left at the table for the dead. Flower-shaped lanterns are sailed across lakes and rivers to help spirits get back home after the celebrations.

Personally, I would feel better in passing knowing I would not be forgotten. My family and I would find comfort in knowing my life, our lives, would be celebrated every year for generations to come. You don't necessarily need to wait for a national holiday to incorporate an annual celebration of your family heritage or to start a meaningful tradition.

One side of my family has *The Quilt*. One person in my family is tasked with sewing a patch onto the quilt when a significant event occurs. When the chosen Quilt timekeeper passes, the quilt is willed to the next person in the generation hierarchy to keep the quilt alive and updated. The quilt serves as a living document of sorts for all the major events in our past. For example, my patch contributions to the quilt are:

- A blanket from when I was born
- The dress from my baptism
- The dress from my wedding

I have also provided patches from all three of my children and their life events. I am not the one that has the quilt, *you are welcome family*, but my cousin is. She dutifully sews on the name, date, event, and patch into the quilt. We are a large family, and it is a massive quilt. If quilts aren't your thing, there are plenty of <u>other</u> traditions and ways to celebrate and keep your heritage intact.

Setting Up Your Family Tradition or Celebration

Ancestry.com makes the money it does because people are interested in where they came from to get in tune with their culture. If you know where you were homegrown, why not look up some customs and traditions from the old country?

For example, a big part of me is Irish. We are big in celebrating Irish Wakes, stopping clocks and covering mirrors at the time of death, and a healthy helping of myth and lore. There are several things that my family could do with this knowledge base.

If we combine everything we know, we could pick St. Patrick Day to celebrate life and death in one swoop. Everyone is a bit Irish on Patty's Day. You can start your journey by doing some light internet research and finding out some fun facts and days of celebration. Simply learning more about where your family has been is honoring them.

Celebrating ancestry allows you to reflect on the past and set new intentions for your future. It is a divine lesson to live life to the fullest, a reset and remember. Have meaningful discussions with family members, swap stories, share emotions, and make memories. Celebrating life and starting a tradition is a reminder to live in the present, count your blessings, and spend quality time with those around you - while you are alive. In my opinion, this is the opposite of losing ourselves - it is the epitome of knowing who we are.

It was said that cemeteries add emotional depth to cities. I think it is the people that do. I agree cemeteries, when maintained and cared for, are beautiful, powerful places of remembrance and interest. However, from everything I have researched, I also believe that cemeteries' continued maintenance has a shelf life. If you think about it, only select cemeteries will be afforded the ability to stand the test of time, the tourist attractions, while others will be lost.

We have already established that we are at capacity for new plots. Alternative methods are being implemented to take over or refurbish the land. Government plans are not something that we can control, but you can sit in the driver's seat with your destiny. Control your final disposition and have 'the talk' with your family.

CHAPTER 6

Having 'The Talk' With Family

Having a body disposition or last wish talk is not an easy topic to breach in certain circles. No one likes to think about their mortality or yours. However, it is not something that should be hurried through at the last minute. Details will be forgotten in a frenzy, and added stress will infest the people you care most about.

Before approaching your family, you may want to have a plan - your plan. After the talk, you will need to make it legal and binding - the earlier, the better. Full-on estate planning is outside the scope of this book; however, what I would suggest is checking out Everplans:

https://www.everplans.com/estate-planning-documents

Everplans has a great book, legal resources, online document storage, and resources for your family members after passing. They offer valuable resources for the executor of your estate. Everplans is a solid place to keep your body disposition instructions and an ideal ice breaker to discuss the topic with your family.

Planning with Everplans ensures there is no guesswork. Pertinent information is organized and available for your last wishes to be implemented. If you approach your family with a well-planned, coherent, thought-out game plan and advise them of the parts they will play, everyone is set for success. Breach the topic over dinner and before you know it, you will be eating pie and laughing at horribly ill-timed jokes in poor taste. One of my favorites:

You can only enter Valhalla if you are in a battle. So, in your last moment, hurl a juice box at the hospice nurse.

It is a precious gift to give your family direction and peace of mind with a plan of action and rules of engagement. Above all, make sure you choose a decent death playlist on your iPad for whatever service you decide to have. No one likes a Debbie Downer.

Remember to be specific and make concrete plans ahead of time with the provider being sure to pay them in advance. You should have the paperwork kept in a safe place like Everplans.

Planning your memorial website is also a breeze with Everplans. They provided their top <u>choices</u>; some are listed here (I used their wording).

Forever Missed

Overview: *Forever Missed is a very user-friendly site. They provide a wide array of templates ranging from military, children's, and everything in between. They provide a limited selection of background music as well. You can log in with Facebook, control the privacy (public or private) and allow visitors to leave virtual flowers and lights. Premium features include unlimited photos and videos, a custom playlist of songs, illustrated stories, and multiple administrators.*

Pricing: *Basic is Free; Monthly: $6, Yearly: $47.88, Lifetime: $95*

GatheringUs

Overview: *GatheringUs is a free comprehensive memorial website that brings communities together after the loss of a loved one. Family and friends can create an online memorial to share memories and photos, post an obituary, designate a nonprofit charity for donations, crowdfund for expenses, send event invites, and track RSVPs. Memorials and event invites can be made public or private, and users can login with Facebook or Google. GatheringUs allows families to find support, celebrate the life of their loved one, and access the necessary resources for their journey.*

Pricing: *Free*

Kudoboard

Overview: *Kudoboard allows you to create a collaborative online memorial that resembles a vibrant and respectful Pinterest board, which can be easily shared with friends and family. Plus, you can download and print a high-resolution version of the board -- or have them print and ship you a poster for a fee -- which can be displayed at a funeral or memorial service to honor the deceased.*

Pricing: *Free for mini memorials (up to 10 posts); $99 one-time charge for a full memorial*

Now For Something Completely Different: EnGraved

I was so intrigued by the conversation with my son that led to this personal discovery journey, I took the disposition of a body a step further. I discussed this journey with my online community in the gaming world and envisioned an app which I named EnGraved and is in the preliminary stages of development and funding at www.vicinanzastudios.com. Nothing like a little shameless promotion (patent pending).

The vision is simple. Engraved is the path to preserving family heritage and human history. It also has the added perk of bringing awareness to sustainability in commemorating loved ones and improving current practices. In writing this book, I found cemeteries are lost (read that again) and we are out of room to bury the dead. Toxins last longer than an eternity. There has to be a better way.

Engraved starts a family and historic preservation effort by equipping gravestones with a QR code embedded with the exact GPS location, video, documents, voice recordings, known family tree, and a link to the departed's 'digital life vault' memorial website. It is our hope that entire cemeteries are fully indexed and public records, archives, vidoes, and more are available to the public.

We encourage people to create their own account on EnGraved, who better to record the story of their life than the person living it. We extend our efforts to equip historical cemeteries with QRs with all known and available information. It is our quest to link the human race to each other and our history through a living index.

I don't want to give it all away, but there is so much this app will be able to do, and I felt it required an honorable-mention in this book. If you would like to contribute to the creation of EnGraved or become a member, please feel free to visit the website.

It is essential we use technology to advocate for a better recording of our lives and making us immortal.

Life After Death

The death of someone you love is the hardest thing to move past. Planning your final disposition with your family is essential in finding inner strength and peace to move on. Celebrating life after death is the hope that carries you forward. Picking your own path adds meaning and value.

Personally, cloning isn't for me (or my animals) because I am a sucker for the imperfections, the nuances, the things that make that person, well, that person.

Having an opportunity to say goodbye and celebrate life as a tribute to your love for the deceased is vital in letting go. Sadly, the family dog is not the only urn we have in the household. Depending on who passes to the unknown first, one of us will be carrying my daughter's ashes with us.

I'm still trying to convince my family that a cherry blossom tree is my wish. It warms my heart thinking my daughter may be with me in nature. But, whatever you decide, money should not be the primary reason you choose it. The choice should align with your beliefs. It's just a body; the soul lives eternally.

My son was not wrong. Investing in a cherry blossom tree provides me peace for an easier transition, as it should for my family. What matters is *how* you spend the time between breaths. And *who* you share breaths with and how you can help your family keep on breathing when your breaths have ceased.

My son's mind is at ease knowing that mom is making it better. My other two children feel comforted they won't have to worry about my urn through every move or garage sale during every generation. To my children, a tree means I am part of something bigger, something beautiful, that will live on. At a minimum, it certainly beats an urn on a dusty old fireplace. You have the opportunity to make your death as meaningful and memorable as your life. Be legendary and immortal - have a care for everyone that comes after you.

SOURCES CITED
BY CHAPTER

Chapter 1:

History, "Human Sacrifice: Why the Aztecs Practiced the Gory Ritual" https://www.history.com/news/aztec-human-sacrifice-religion

Neptune Society, "Hinduism and Creation" https://www.neptunesociety.com/cremation-information-articles/hinduism-and-cremation

NPR, "When People Ate People a Strange Disease Emerged" https://www.npr.org/sections/thesalt/2016/09/06/482952588/when-people-ate-people-a-strange-disease-emerged

PBS, "Global Connections" http://www.pbs.org/wgbh/globalconnections/mideast/themes/religion/index.html

The History of Indian Funerals, "Asian Funeral History" https:// thefuneralsource.org/hi0204.html

Wikipedia, "Burial" https://en.wikipedia.org/wiki/Burial

Chapter 2:

Considerable, "Death Is Forever. Cemeteries as They Currently Exist, Might Not Be" https://www.considerable.com/life/death/ future-cemeteries-burial-space/

Discover, "The Taj Mahal: Can India Save This Corroding Beauty?" https://www.discovermagazine.com/planet-earth/ the-taj-mahal-can-india-save-this-corroding-beauty

India Today, "Agra: Taj Mahal to undergo major security upgrade next year" https://www.indiatoday.in/cities/agra/ story/agra-taj-mahal-to-undergo-major-security-upgrade-next-year-1817064-2021-06-20

Insider, "Traditional Burials Are Ruining the Planet - Here's What We Should Do Instead" https://www.businessinsider.com/ traditional-burials-are-ruining-the-planet-2016-4

Medium, "The Taj Mahal - under Environmental and Political Threat" https://medium.com/the-naked-architect/

the-taj-mahal-under-environmental-and-political-threat-7b3b8c4b7f6f

Mental Floss, "8 Cemeteries Unearthed at Construction Sites" https://www.mentalfloss.com/article/535662/cemeteries-unearthed-at-construction-sites

Tampa Bay Times, "In Search of Lost Cemeteries" https://projects.tampabay.com/projects/2019/special-reports/missing-tampa-cemeteries-map/

Vice, "Your Favorite Park is Probably Built On Dead Bodies" https://www.vice.com/en/article/akwp8e/your-favorite-park-is-probably-built-on-dead-bodies

Wikipedia, "Cremation" https://en.wikipedia.org/wiki/Cremation

Chapter 3:

Barton Family Funeral Services, "Embalming History" https://bartonfuneral.com/funeral-basics/history-of-embalming/

Cremation Solutions, "The Horrible Effects of Formaldehyde on Funeral Directors" https://www.cremationsolutions.

com/blog/the-horrible-effects-of-formaldehyde-on-funeral-directors/2017/02/

Medical News Today, "What Is Arsenic Poisoning?" https://www.medicalnewstoday.com/articles/241860

New York Times, "Despite Cancer Risk, Embalmers Still Embrace Preservative" https://www.nytimes.com/2011/07/21/business/despite-cancer-risk-embalmers-stay-with-formaldehyde.html

Smithsonian, "Arsenic and Old Graves: Civil War-Era Cemeteries May Be Leaking Toxins" https://www.smithsonianmag.com/science-nature/arsenic-and-old-graves-civil-war-era-cemeteries-may-be-leaking-toxins-180957115/

Smithsonian, "Egyptian Mummies" https://www.si.edu/spotlight/ancient-egypt/mummies

Smithsonian, "When You Die You Will Probably Be Embalmed. Thank Lincoln For That" https://www.smithsonianmag.com/science-nature/how-lincolns-embrace-embalming-birthed-american-funeral-industry-180967038/

The Guardian, "Should I Be Buried or Cremated" https://www.
theguardian.com/environment/2005/oct/18/ethicalmoney.
climatechange

USGS, "The Water In You: Water and the Human Body"
https://www.usgs.gov/special-topic/water-science-school/
science/water-you-water-and-human-body?qt-science_center_
objects=0#qt-science_center_objects

Vice, "Flooded Corpses Are Leaking Formaldehyde into
Northern Ireland's Groundwater" https://www.vice.com/en/
article/qbx8ed/worse-than-zombies-246

Wikipedia, "Mummy" https://en.wikipedia.org/wiki/Mummy

Chapter 4:

Alcor https://www.alcor.org/what-is-cryonics/

Anatomy Gifts Registry http://www.anatomygifts.org/

BioUrns https://urnabios.com/urn/

Capsula Mundi https://www.capsulamundi.it/en/

Celestis https://www.celestis.com/
launch-schedule/#upcoming-launches

Eternal Reefs https://www.eternalreefs.com/

Eterneva https://eterneva.com/loved-ones

Green Burial Council https://www.greenburialcouncil.org/why_certification_matters.html

Heavenly Stars Fireworks http://www.heavenlystarsfireworks.com/

Interactive Estate Document Systems https://ieds.online/body-donation-programs-by-state/

Medcure https://medcure.org/

News Medical, "What Happens to the Brain After Death" https://www.news-medical.net/health/What-Happens-to-the-Brain-After-Death.aspx

Physicians Committee, "Donate Your Body to Science" https://www.pcrm.org/ethical-science/animal-testing-and-alternatives/donate-your-body-to-science

Research for Life https://www.researchforlife.org/

ScienceCare https://www.sciencecare.com/about-science-care

Sinogene https://www.sinogene.org/pet-cat-cloning.html

Today I Found Out, "What Happens When You Donate Your Body to Science and How You Do This" http://www.todayifoundout.com/index.php/2018/01/donate-body-science-death/

United Tissue Network https://unitedtissue.org/

Variety http://variety.com/2018/film/news/barbra-streisand-oscars-sexism-in-hollywood-clone-dogs-1202710585/

Chapter 5:

Body Worlds Exhibit https://bodyworlds.com/

Considerable, "More Americans are Skipping Traditional Funerals in Favor of Green Burials" https://www.considerable.com/life/communication/green-burials-are-a-growing-trend/

Culture Trip, "10 Ways to Honor the Dead Around the World" https://theculturetrip.com/africa/articles/10-ways-to-honor-the-dead-around-the-world/

Genealogy Pals, "How Do You Honor Ancestors? 12 Family History Celebrations" https://genealogypals.com/how-do-you-honor-ancestors-12-family-history-celebrations/

Mutter Museum http://muttermuseum.org/news/podcasts/

Walks of Italy, "The Unbelievable Story of the Paris Catacombs" https://www.walksofitaly.com/blog/art-culture/paris-catacombs

Chapter 6:

Forever Missed http://forevermissed.com/

GatheringUs https://www.gatheringus.com/

Kudoboard https://www.kudoboard.com/online-memorial

Printed in the United States
by Baker & Taylor Publisher Services

Printed in the United States
by Baker & Taylor Publisher Services